DID YOU KNOW?
ANIMAL ADAPTATIONS



20 THINGS YOU DIDN'T KNOW ABOUT

# REPTILE ADAPTATIONS

SLOANE HUGHES

PowerKiDS press
New York

Published in 2023 by The Rosen Publishing Group, Inc.
29 East 21st Street, New York, NY 10010



Portions of this work were originally authored by Barbara M. Linde and published as *20 Fun Facts About Reptile Adaptations.* All new material in this edition was authored by Sloane Hughes.

Editor: Amanda Vink
Book Design: Tanya Dellaccio

Photo Credits: Cover (speech bubble) Gohsantosa/Shutterstock.com; cover (logo art) NazArt/Shutterstock.com; Cover Lauren Suryanata/Shutterstock.com; p. 5 (snake) Kurit afshen/Shutterstock.com; p. 5 (chameleon) Ondrej Prosicky/Shutterstock.com; p. 5 (turtle) Sakis Lazarides/Shutterstock.com; p. 5 (iguana) Enessa Varnaeva/Shutterstock.com; p. 6 Jason Mintzer/Shutterstock.com; p. 7 taviphoto/Shutterstock.com; p 8 Mark_Kostich/Shutterstock.com; p. 9 tsnebula23/Shutterstock.com; p. 10 meunierd/Shutterstock.com; p. 11 (alligator) Eric Isselee/Shutterstock.com; p. 11 (crocodile) Pumbastyle/Shutterstock.com; p. 12 GUDKOV ANDREY/Shutterstock.com; p. 13 Sista Vongjintanaruks/Shutterstock.com; p. 14 NFKenyon/Shutterstock.com; p. 15 Jay Ondreicka/Shutterstock.com; p. 16 Shane Myers Photography/Shutterstock.com; p. 17 OrodO/Shutterstock.com; p. 18 ilipap/Shutterstock.com; p. 19 (milk snake) Eric Isselee/Shutterstock.com; p. 19 (coral snake) Jay Ondreicka/Shutterstock.com; P. 20 Mike Wilhelm/Shutterstock.com; p. 21 Matt Jeppson/Shutterstock.com; p. 22 V.P.I/Shutterstock.com; p. 23 bluefish_ds/Shutterstock.com; p. 25 Sergey Uryadnikov/Shutterstock.com; p. 26 Natalia Kuzmina/Shutterstock.com; p. 27 woodphotography LLC/Shutterstock.com; p. 29 Sandra Foyt/Shutterstock.com.

**Library of Congress Cataloging-in-Publication Data**
Names: Hughes, Sloane, author.
Title: 20 things you didn't know about reptile adaptations / Sloane Hughes.
Other titles: Twenty things you didn't know about reptile adaptations
Description: New York : PowerKids Press, [2023] | Series: Did you know? animal adaptations | Includes index.
Identifiers: LCCN 2021061798 (print) | LCCN 2021061799 (ebook) | ISBN 9781538386873 (library binding) | ISBN 9781538386859 (paperback) | ISBN 9781538386866 | ISBN 9781538386880 (ebook)
Subjects: LCSH: Reptiles--Adaptation--Juvenile literature.
Classification: LCC QL644.2 .H85 2023 (print) | LCC QL644.2 (ebook) | DDC 597.914--dc23/eng/20220106
LC record available at https://lccn.loc.gov/2021061798
LC ebook record available at https://lccn.loc.gov/2021061799

Manufactured in the United States of America

CPSIA Compliance Information: Batch #CSPK23. For Further Information contact Rosen Publishing, New York, New York at 1-800-237-9932.

# CONTENTS

# REMARKABLE REPTILES

What makes a reptile a reptile? Reptiles are invertebrates, which means they have backbones. They're cold-blooded. They use their environment, or surroundings, to control their body heat. They're also covered in scales. Some lay eggs, and some give birth to live young.

Reptiles have special features, or adaptations, that help them survive and thrive on every continent except Antarctica. Let's explore these **unique** reptile adaptations!

IGUANA

The earliest reptiles date back to more than 300 million years ago!

GREEN SEA TURTLE

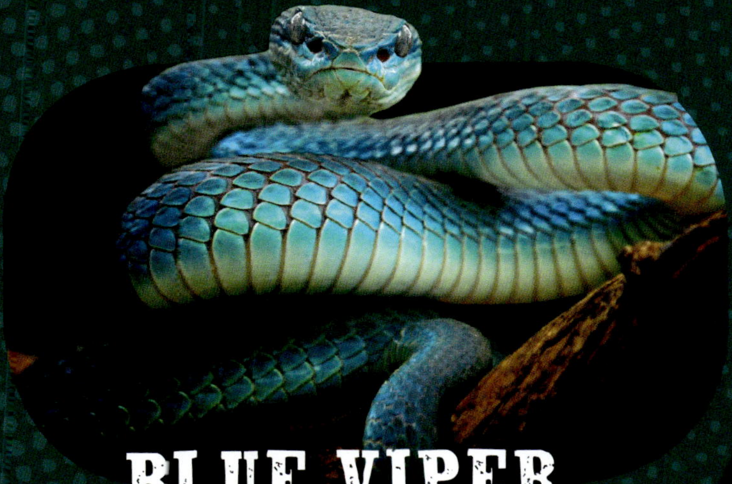

BLUE VIPER SNAKE

CHAMELEON

# SKIN DEEP

**DID YOU KNOW?**

Reptiles depend on the sun to maintain body heat.

A reptile can't keep its body warm with its own energy. Instead, it has to find an outside source of warmth. When reptiles are cold, they bask, or lay in the sun to warm up.

The term "cold-blooded" doesn't mean that a reptile has icy blood, just that it can't keep itself warm without outside help. This desert iguana works to keep its body temperature around 101.3 degrees F (38.5 degrees C).

**DID YOU KNOW?**

Scales act as reptile armor, but they also keep a reptile's body from drying out.

Reptile scales are made of a strong matter called keratin. This is the same material in our hair and nails. Scales **overlap** in rows and give a snake or lizard **protection**. They're also very tough and waterproof. Snakes and lizards can live away from

DID YOU KNOW?

All reptiles shed their skin as they grow larger.

A reptile's skin doesn't grow as it grows. When they get too big for their skin, reptiles need to molt, or shed their old skin. This molting is different for different species. Turtles shed in pieces, while snakes shed their skin all at once. They crawl out of it, nose first!

A snake's body returns to normal once it's done **digesting** its meal. That can take a while!

**DID YOU KNOW?**

A snake's body can **stretch** a huge amount!

A snake's jaws open really wide! Not only that, but a snake's skin stretches and its bones can move around. These adaptations let a snake eat prey that is much bigger that it is!

# TEETH AND TONGUES

DID YOU KNOW?

Alligators and crocodiles never stop growing teeth.

Sharp teeth are very important for members of the animal order Crocodilian. That's why they keep growing new ones. An alligator, for example, may grow as many as 3,000 teeth over the course of its life!

## TO TELL THE DIFFERENCE BETWEEN ALLIGATORS AND CROCODILES, LOOK AT THEIR TEETH!

### CROCODILE

- POINTED, V-SHAPED SNOUT
- SAME SIZED UPPER AND LOWER JAWS
- TEETH STICK OUT WHEN MOUTH IS CLOSED

Mature, or grown-up, alligators have about 80 teeth in their mouth at a time.

### ALLIGATOR

- WIDE, ROUNDED, U-SHAPED SNOUT
- WIDE UPPER JAW
- TEETH DON'T SHOW WHEN MOUTH IS CLOSED

In fact, it's longer than the chameleon's body. The chameleon is really fast at using its tongue too, and the end of the tongue is sticky and shaped like a club. Insects beware!

**DID YOU KNOW?**

A chameleon has a really, *really* long tongue.

Alligator snapping turtles can hold their breath underwater for 40 to 50 minutes!

DID YOU KNOW?

The alligator snapping turtle uses its tongue to go fishing.

When the alligator snapping turtle is on the hunt, it keeps most of its body still. But it does open its mouth and move its tongue. Its tongue has a bright red tip that looks like a worm. This can draw an **unsuspecting** fish in close. Then, snap! Dinner!

# REPTILIAN TRICKS

**DID YOU KNOW?**

Some alligators and crocodiles set a trap to trick water birds into coming close to them.

A clever alligator or crocodile might put sticks on its snout. Water birds are on the lookout for materials to build their nests. When the birds fly down to get the sticks, they can become a quick snack!

Wood turtles may also use their chest to bump the ground when searching for worms.

**DID YOU KNOW?**

Wood turtles have learned to dance for their food.

When they're hungry, wood turtles may begin to rock back and forth, stomping their feet on the ground. They gobble up worms coming to the surface because of the movement.

# BLENDING IN

**DID YOU KNOW?**

The special coloring of green sea turtles helps them blend in from above and below.

Green sea turtles are dark brown or green on top. This helps them blend in with the darkness of the water below them. The underside of their body is lighter. That helps them blend in with the world above when an animal looks up at them from below.

The eastern green mamba snake can bite quickly when **threatened**, and it will often do so multiple times. Its venom can cause death quickly.

**DID YOU KNOW?**

Green mamba snakes could win a game of hide and seek!

When an animal blends in with its surroundings, that's called camouflage. Eastern green mamba snakes look a lot like the plants they rest in. They are shy creatures, but when they are bothered, watch out! They're **venomous**!

**DID YOU KNOW?**

Chameleons can change color using special crystals in their skin.

Chameleons sometimes change color, but it's not to blend in. These color changes are sometimes caused by changes in light or temperature. They might show a chameleon's mood. A male chameleon might change color when it is scared or when it has won a fight.

A chameleon's natural coloring allows it to blend in with its surroundings.

CORAL SNAKE

MILK SNAKE

The milk snake is not venomous. However, it has similiar colors to the coral snake, which is highly venomous. Predators prefer to avoid both milk snakes and coral snakes!

DID YOU KNOW?

The harmless milk snake has adapted to **mimic** a deadly snake so **predators** stay away.

# STAYING SAFE

**DID YOU KNOW?**

The hognose snake has several tricks to avoid predators, from acting tough to acting dead!

If the hognose snake meets a predator, the first thing it will do is puff up and flatten its head. That makes it look scary. It will also hiss loudly and pretend to strike. If none of that works, it rolls over, opens its mouth, and lets its tongue hang out.

If a predator grabs a gecko and it can't escape, the gecko still has one last trick. Its tail has special places where it can break off. The confused predator watches the tail moving on the ground, letting the lizard run to safety.

**DID YOU KNOW?**

A gecko's tail can break off. It continues to move while the rest of the gecko escapes.

**DID YOU KNOW?**

Reptile eggshells are **leathery** and hard, and they don't break easily.

This gives reptiles more choices about where to make their nests and lay their eggs. Many reptile babies use a special egg tooth to poke a hole in the shell when they're ready to hatch.

Sand helps the eggs stay at the proper temperature and keeps them from drying out.

**DID YOU KNOW?**

Female sea turtles lay their eggs on the same beach on which they were born.

Female sea turtles may travel hundreds of miles to lay their eggs in a hole in the sand—and then they leave! Sea turtle babies hatch together, and they move as quickly as possible toward the sea. That way, more of them may be able to escape predators.

# DEADLY PREDATORS

**DID YOU KNOW?**

People once thought a Komodo dragon's mouth was so dirty that bacteria killed its prey.

For a long time, people thought that a bite from a Komodo dragon would cause its prey to die because of the deadly bacteria in the lizard's mouth. In fact, a Komodo dragon's mouth is no dirtier than that of any other lizard. Scientists have discovered, however, that a Komodo dragon's bite contains venom.

Komodo dragons are the largest lizards on Earth. Adults can reach 10 feet (3 m) long and weigh more than 300 pounds (140 kg)!

## SOME REPTILES ARE VERY SMALL, AND OTHERS ARE REALLY LONG!

### ADULT AVERAGE LENGTH OF REPTILES

NANO-CHAMELEON: 0.5 INCH (1.3 CM) LONG
CHAMELEON: 15 TO 20 INCHES LONG (38.1 TO 50.8CM)
PREHENSILE-TAILED SKINK: 30 INCHES (76 CM) LONG
SEA TURTLE: 3 TO 4 FEET (83 TO 114 CM) LONG
KOMODO DRAGON: 10 FEET (3 M) LONG
ANACONDA: 30 FEET (9 M) LONG

KOMODO DRAGON

**DID YOU KNOW?**

Boa constrictors hug their prey to death!

Boa constrictors are not venomous. Instead, they wrap themselves tightly around their prey and constrict, or squeeze really tight. This stops blood from getting to the prey's brain, and it dies soon after.

The boa constrictor swallows its prey whole!

DID YOU KNOW?

A baby rattlesnake doesn't get a rattle until it sheds its skin the first time.

Baby rattlesnakes are born with a "first button." This is a hard piece of keratin on the end of their tail. Every time a rattlesnake sheds its skin, a new "button" is created. Rattlesnakes use these hollow buttons to make noise to scare away predators.

# RAVING ABOUT REPTILES

Reptiles are amazing creatures! Go almost anywhere in the world, and you'll find a wide variety of these animals.

If you want to learn more about reptiles, visit a natural history museum or a zoo. Talk to a herpetologist. This is a person who studies reptiles. They'll be able to tell you a lot about reptiles and their amazing adaptations! They may even show you how to properly handle a reptile.

parks

# GLOSSARY

**digest:** To break down food inside the body so that the body can use it.

**leathery:** Looking or feeling like leather.

**mimic:** To copy the way something looks, acts, or sounds.

**overlap:** To place or be placed so that a part of one thing covers a part of another.

**predator:** An animal that hunts other animals for food.

**protection:** The act of keeping something safe from harm.

**stretch:** To extend the body or limbs.

**threatened:** Likely to be harmed.

**unique:** Special or different from anything else.

**unsuspecting:** Unaware of any danger or threat.

**venomous:** Having a poisonous bite or sting.

# FOR MORE INFORMATION

## BOOKS

Csotonyi, Julius T. *Discovering Reptiles: The Ultimate Handbook to the Reptiles of the World!* New York, NY: Simon & Schuster, 2021.

Feldman, Thea. *Alligators and Crocodiles Can't Chew! And Other Amazing Facts*. New York, NY: Simon & Schuster, 2021.

## WEBSITES

**American Museum of Natural History**

*www.amnh.org/explore/ology/zoology/meet-the-ologist-chris-raxworthy*

Want to meet a herpetologist, or a person who studies reptiles? Watch a video on this website!

**National Geographic Kids: Reptiles**

*kids.nationalgeographic.com/animals/reptiles*

Is there a reptile you want to know more about? This website will give you information about specific species.

**Publisher's note to educators and parents:** Our editors have carefully reviewed these websites to ensure that they are suitable for students. Many websites change frequently, however, and we cannot guarantee that a site's future contents will continue to meet our high standards of quality and educational value. Be advised that students should be closely supervised whenever they access the internet.

# INDEX